Cambridge checkpoint

Cambridge Assessment
International Education
Endorsed for learner support

Lower Secondary Science

WORKBOOK

8

Peter D Riley

Boost

HODDER
EDUCATION

Cambridge International copyright material in this publication is reproduced under licence and remains the intellectual property of Cambridge Assessment International Education.

Photo credits

p.2 *l* © Anton/stock.adobe.com; **p.2** *r* © Eartty/stock.adobe.com

Every effort has been made to trace all copyright holders, but if any have been inadvertently overlooked, the Publishers will be pleased to make the necessary arrangements at the first opportunity.

Although every effort has been made to ensure that website addresses are correct at time of going to press, Hodder Education cannot be held responsible for the content of any website mentioned in this book. It is sometimes possible to find a relocated web page by typing in the address of the home page for a website in the URL window of your browser.

Third-party websites and resources referred to in this publication have not been endorsed by Cambridge Assessment International Education.

Hachette UK's policy is to use papers that are natural, renewable and recyclable products and made from wood grown in well-managed forests and other controlled sources. The logging and manufacturing processes are expected to conform to the environmental regulations of the country of origin.

Orders: please contact Hachette UK Distribution, Hely Hutchinson Centre, Milton Road, Didcot, Oxfordshire, OX11 7HH. Telephone: +44 (0)1235 827827. Email: education@hachette.co.uk Lines are open from 9 a.m. to 5 p.m., Monday to Friday. You can also order through our website: www.hoddereducation.com

ISBN: 978 1 3983 0141 2

© Peter D Riley Ltd 2022

First published in 2012

This edition published in 2022 by
Hodder Education,
An Hachette UK Company
Carmelite House
50 Victoria Embankment
London EC4Y 0DZ

www.hoddereducation.com

Impression number   10 9 8 7 6 5 4

Year   2026 2025 2024 2023

Cover photo © Satit_Srihin – stock.adobe.com

Illustrations by Integra Software Services Pvt. Ltd., Pondicherry, India

Typeset in Integra Software Services Pvt. Ltd., Pondicherry, India

Printed in the UK

A catalogue record for this title is available from the British Library.

Contents

Introduction

The aim of every science course is to help you become scientifically literate or, more simply, to help you become a 'scientific citizen'. This means that you can confidently talk and write about the science you have studied and know how it helps us to understand and live in our world. Below are some questions that a scientific citizen should be able to answer. Just read through them slowly.

Name two places where you can find a hinge joint in your body. What is the shape of a red blood cell? When you breathe in, do your ribs move up or down? What food group does sugar belong to? Why do smokers develop a cough? What is bioaccumulation and why is it harmful? Where in an atom do you find protons and neutrons? What is a pure substance? List the top five metals of the reactivity series in the correct order. What is oxidation? What does 'accelerate' mean? What kind of force occurs when one object crashes into another? Define the term 'weight'. Name five luminous objects. Name three types of simple machine. What is the difference between weather and climate? What is a rogue planet? How are scientific enquiries carried out and how can you find out information from them?

The chances are that you will not be able to answer most of those questions now, but if you work through this book as you complete the chapters in the *Checkpoint Science Stage 8 Student's Book*, you will be on your way to being scientifically literate.

Here is the first challenge of the book – look at these questions again and write down any answers you might have. It does not matter if you cannot think of any answers to a question, just keep a record of your answers for later. When you have completed this workbook, look back at these questions again, write down your answers and see how they differ from your first answers. This should show you that you are well on your way to being a scientific citizen.

Most of the questions aim to test your knowledge and understanding of science, but some questions have this icon ⭐. These questions aim to test your science enquiry skills.

Some other questions have this icon 🔵. These are questions about using models to help you learn about and understand scientific ideas.

Yet other questions have this icon 🧩. These questions put science in context.

So now it is time to start. Read each question carefully, think about it, then write down the answer in the space provided, or as otherwise instructed. Often you will have to write down some facts or explanations, sometimes you will need to tick a box and occasionally you will have to link details by lines or construct and interpret graphs.

BIOLOGY

1 Joints and muscles

Joints

1 What is a joint? Tick (✔) **one** box.

A place where a bone meets a muscle. ☐

A place where two bones meet. ☐

A place where two muscles meet. ☐

A place where three muscles meet. ☐

2 **a** Name **two** places in the body where you can find a hinge joint.

   ....

   b Name **two** places in the body where you can find a ball and socket joint.

   ....

3 **a** What kind of joint allows part of a limb (an arm or a leg) to move backwards and forwards?

   ....

   b What kind of joint allows the arm or leg to move backwards, forwards *and* from side to side?

   ....

4

   a Where in the body is this joint found? ....

   b What kind of joint is it? ....

5

A

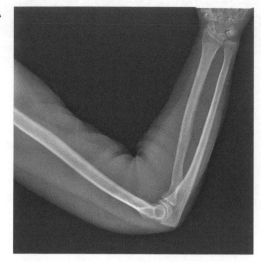

B

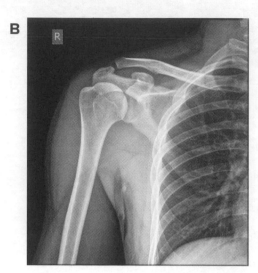

a What type of joint is shown in A?

....

b Where in the body is the joint shown in A?

....

c What type of joint is shown in B?

....

d Where in the body is the joint shown in B?

....

A closer look at the muscular system

6 Which of the following statements about muscles are correct? Tick (✔) **two** boxes.

Muscles pull on bones when they increase in length. ☐

Muscles pull on bones when they decrease in length. ☐

A muscle can increase and decrease in length on its own. ☐

Muscles shorten when they contract to move a bone. ☐

When systems work together

7

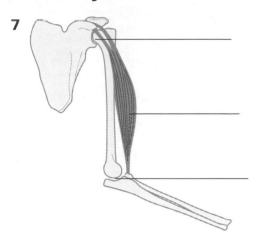

a On the diagram above, label

   i a hinge joint   **ii** a ball and socket joint   **iii** the muscle.

b Which part of the body in the diagram provides the power for movement?

....

c When this part provides the power for movement, describe how it changes.

....

....

....

d When this part moves,

   i which bone or bones move?

....

   ii Draw an arrow to show the direction of movement of the bone or bones.

8

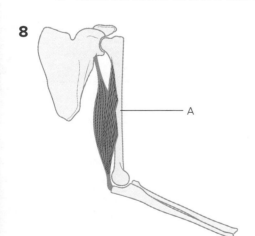

a What is the name of the part labelled A? ....

b Draw an arrow on the diagram to show the direction of the movement when this part provides the power to move.

9 You have been asked to make a model of the bones and both muscles in the arm.

a What materials or objects would you use for the bones?

....

b What materials or objects would you use for the muscles? They must be able to change their length as you demonstrate how they provide movement.

....

c Draw a diagram of your model in the space provided.

10 The biceps and triceps are called an **antagonistic muscle pair**. What does this mean?

....

....

....

2 Blood

Red blood cells

1 The figure below shows a sample of blood which has been separated into three parts. Name the **two** parts marked A and B.

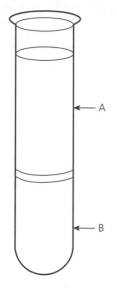

A ....

B ....

2 Describe the shape of a red blood cell.

....

....

....

3 What is the function of a red blood cell?

....

4 Red cells are slightly purple when blood moves towards the lungs. When they move away from the lungs they are bright red. Explain the reason for this change.

....

....

....

....

White blood cells

5 Red blood cells and white blood cells are different in colour. State **two** other ways in which they are different.

....

....

6 What are disease-causing microorganisms called? ....

7 If a person suffers from sickle cell anemia,

   a what type of cells are affected?

....

   b what are the shape of the cells that are affected by the disorder?

....

   c why are these cells dangerous?

....

   d what does the patient take to control and ease the condition?

....

   e on which continent are they most likely to live?

....

8 a What is a pathogen?

....

   b Name **two** examples of pathogens.

....

....

c Name **three** places in the body where pathogens may enter.

....

....

....

9 a What type of cell attacks pathogens?

....

b How does the cell attack the pathogen?

....

c What happens to the cell when it has finished its attack?

....

Plasma

10 a What is plasma?

....

....

b Name **three** things transported in the plasma.

....

....

....

11 You are to make a model of blood.

a What would you use as an analogy for

i red cells? ....

ii white cells? ....

iii plasma? ....

b In your model, which arrangement of cells would you use to show their numbers in real blood? Underline the correct answer.

    A A lot more red cells than white cells.

    B About the same number of red cells and white cells.

    C A lot more white cells than red cells.

c In the space provided below, draw a diagram of how you would set up your model.

d Label the parts that you show on your diagram.

12 a Who is a blood donor?

....

....

b Why do we need blood donors?

....

....

c What happens in a blood transfusion in a hospital?

....

....

3 The respiratory system

Parts of the respiratory system

1 a Use these words to write a summary word equation for aerobic respiration.

Water, oxygen, carbon dioxide, glucose

....

b Where precisely does aerobic respiration take place in the bodies of plants and animals?

....

2 What does the term 'breathing' mean?

....

3

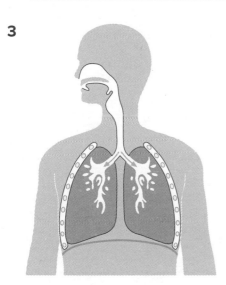

a The diagram above shows the structure of the respiratory system. Label all of the parts you know.

b For each part that you know, explain what its purpose is in the respiratory system.

....

....

....

....

....

....

....

....

4 **a** Name **three** substances that can cause some people to have an asthma attack.

....

....

....

b If a person is allergic to these substances what does it mean?

....

....

c How does an asthma attack affect a person's breathing?

....

....

d When someone has an asthma attack they may use a device to help them.

i What is the name of this device which begins with the letter i?

....

ii How does this device help?

....

Breathing movements

5 The information in the left-hand column in the table on the next page describes the movements that occur during breathing. The other two columns contain boxes for you to tick. For each breathing movement, put a tick (✔) in **one** of the two columns to show whether the movement happens during inspiration or expiration.

Breathing movement	Inspiration	Expiration
external intercostal muscles relax		
ribs move up		
diaphragm muscles contract		
chest volume decreases		
ribs move down		
air moves out		
air moves in		
diaphragm muscles relax		
chest volume increases		
external intercostal muscles contract		

6 Aya tested the breathing rate of Jin for three activities and recorded them in a table.

Activity	Trial 1 Breaths per minute	Trial 2 Breaths per minute	Trial 3 Breaths per minute	Trial 4 Breaths per minute	Trial 5 Breaths per minute
Rest	15	16	15	14	17
After walking	25	11	27	25	24
After running	45	48	44	45	46

a How did Aya attempt to make her data reliable?

....

....

b Is there a trend in the data? Explain your answer.

....

....

....

c Is there an anomalous result? Explain your answer.

....

....

....

d What conclusion can you draw from Aya's data?

....

....

....

e What are the limitations of the conclusion?

....

....

....

f How could the investigation be improved?

....

....

7 Jin believes that when his kitten is at rest, its breathing rate is less than when it has been running.

Is this hypothesis testable? Explain your answer.

....

....

....

....

Exchanging gases

8 The diagram below shows an air sac and the movement of gases in it.

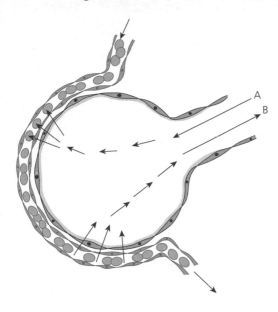

a Where in the body would you find an air sac like this?

....

b What is the name of the tube running around the side of the air sac?

....

c What are shown in this tube?

....

d What is the name of the gas which moves in direction A?

....

e What is the name of the gas which moves in direction B?

....

f What is the process called by which the gases move across the air sac?

....

A healthy diet

Nutrients needed by the body

1 a Name **two** substances which are carbohydrates.

....

....

b Why are carbohydrates important in the diet?

....

....

2 a Name a liquid fat produced in plants.

....

b Name **three** ways that the body uses fats.

....

....

....

3 a Name **two** structures that proteins form in the body.

....

....

b Give **two** other reasons why the body needs proteins.

....

....

How the body uses nutrients

4 Match each of the vitamins listed below to their effect on the body, and choose a good source for each by drawing lines between the boxes.

Effect on body	Vitamin	Good sources
prevents rickets	A	blackcurrant and guava
increased resistance to disease	C	egg yolk and sunlight
prevents scurvy	D	milk and liver

5 a Name **two** structures in the body that require calcium.

....

b Which cells in the body particularly need iron?

....

c How does the presence of iron in these cells help them perform their task in the body?

....

....

d If the diet does not supply enough iron to the body, what condition can develop?

....

6 a What percentage of a person's body weight is due to water in the body?

....

b Why does the body need water?

....

7 A scientist performed experiments that showed a food helped to cure an illness. She sent her data and conclusions for peer review.

    a What does peer review mean?

....

....

....

....

    b How does peer review help our understanding of science?

....

....

....

....

The nutrients in food and Energy in food

8 The table below shows the nutrients in a few common foods.

Food/100g	Protein g	Fat g	Carbohydrate g	Energy/kJ
Rice	6.2	1.0	86.8	1531
Lentil	23.8	0	53.2	1256
Chicken	20.8	6.7	0	602
Peanut	28.1	49.0	8.6	2428

a For each of the nutrients listed below, put the foods shown in the table in order, starting with the highest value of each nutrient, and ending with the lowest value of each nutrient.

i protein

....

....

ii fat

....

....

iii carbohydrate

....

....

iv energy.

....

....

b Compare the fat and carbohydrate values for rice and peanuts with their energy values. Which nutrient, fat or carbohydrate contains the greater amount of energy in each gram of substance?

....

....

....

9 Why is a balanced diet important for

a growth of the body?

....

....

b development of the body?

....

....

c health of the body?

....

....

 10 In the early part of the twentieth century, food scientists identified that food contained proteins, fats, carbohydrates and minerals, and believed that these were the only substances needed for a healthy diet.

Frederick Gowland Hopkins knew that all young mammals were fed on milk by their mothers, and reasoned that this must be the best food for them. However, if the four nutrients were the only substances needed for animals to grow well, the addition of milk to the diet should make no difference to the growth of young animals.

To test his idea, he set up the following investigation. He took a group of rats (Group 1), weighed them to find their average weight and fed them on food made from the four nutrients. He also weighed them at regular intervals.

He took a second group (Group 2) and treated the rats like those in the first group except that he gave them a small amount of milk with their food. Here is a table of his results.

Day	Group 1 Average weight of rats/g	Group 2 Average weight of rats/g
0	45	45
3	47	51
6	50	58
9	51	63
12	51	68
15	47	71

a Plot the data on the graph below. You should make two line graphs.

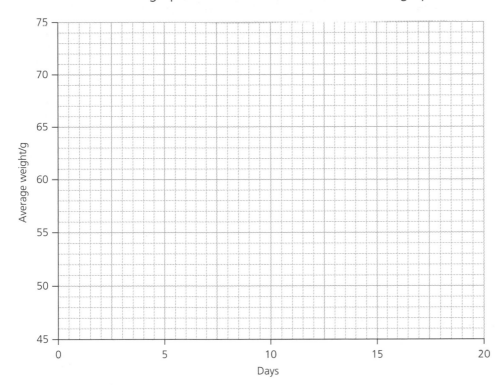

b What does the graph of the average weight of rats in Group 1 show?

....

....

c What does the graph of the average weight of rats in Group 2 show?

....

....

d How did the addition of milk to their diet affect the growth of the rats?

....

....

e How did this result compare with the ideas of the other food scientists?

....

....

f What previous knowledge did Hopkins use to set up his investigation?

....

....

g Hopkins took his investigation further by removing milk from the diet of the rats in Group 2 and giving it to the rats in Group 1. What do you predict happened to the rats in

i Group 1? ....

ii Group 2? ....

A healthy diet

11 Draw a triangle in the space below and, at its peak, write the names of foods that you should only eat in small amounts. At its base, write the names of foods that you can eat in large amounts.

5 A healthy lifestyle

Diet

1 a The Russian scientist Nikolay Anichkov studied the effect of a high cholesterol diet on rabbits. What did he discover?

....

....

b When his work was eventually studied by other scientists, they carried out tests on people in seven countries across the world.

What was the advantage of making such a study?

....

....

c Why is it important to keep the amounts of cholesterol in the diet low?

....

....

d Name **three** types of food that should be eaten in small amounts, in order to stop cholesterol building up in the body.

....

....

e Name **four** types of food which can be eaten to avoid cholesterol building up in the body.

....

....

2 Why can people who lack iron in their diet feel permanently tired?

....

....

3 How can a person become obese?

....

....

....

....

....

....

....

....

4 a What science topics does a dietitian study to a high level?

....

....

b Where does a dietitian work?

....

c Who can a dietitian help?

....

A healthy heart

5 The heart can be damaged by an unbalanced diet and smoking. Why do people need to have a lifestyle that keeps their heart healthy?

....

....

....

6 The condition of the heart and the effect of lifestyle on it can be checked by measuring the pulse.

a What is the pulse and what does it tell you about the heart?

....

....

....

....

....

b How would you instruct someone to measure their pulse?

....

....

....

....

....

7 The condition of the heart as someone performs various activities can be checked by taking the pulse with a heart monitor. The results can be used to check the health of the heart to see if it is affected by lifestyle.

Su Lin is wearing a heart monitor that records her pulse. She sits down to wait for a bus. She stands up when she sees the bus coming. The bus goes past her, then stops and she runs after it. She gets on the bus, then sits down. The table below shows the readings of her heart monitor during this time.

Time/seconds	Pulse rate/beats per minute
0	70
30	76
60	76
90	120
120	134
150	110
180	90
210	76
240	70

a Make a line graph from the data in the table.

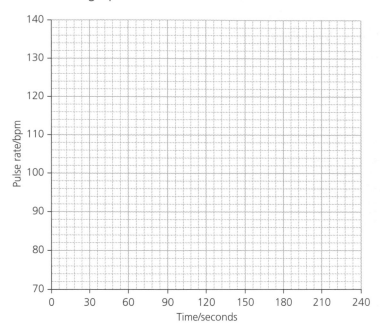

b What was Su Lin's pulse rate when she was sitting down? ....

c What was Su Lin's pulse rate when she stood up? ....

d Identify the time period when Su Lin was running for the bus.

....

e How long did it take Su Lin's heart to recover from running?

....

f The fitness of the heart can be assessed by measuring the time it takes to recover from exercise. The fitter the heart, the faster it recovers.

Three other students recorded the time it took them to recover from exercise. Their results are in the table below.

Name	Recovery time/minutes
Carlos	2.5
Adya	1.95
Clare	3.25

Compare Su Lin's results with the other students', then write their names in the order of fitness, starting with the person who has the fittest heart.

....

....

Exercise and the heart

8 Scientists advise people to take exercise as part of a healthy lifestyle. State **two** ways in which regular exercise helps the body to stay healthy.

....

....

....

....

9 Sports activities can be a useful part of a healthy lifestyle, but many people stop taking part in sports when they leave school. What advice would you give to someone who has not been active for a while, but has decided to take up sports activities again?

....

....

....

....

Smoking and health

10 a Name **two** ways in which smoking can damage the body.

....

....

b Will smoking affect the growth, development and health of the body?

....

....

....

6   Ecosystems around the world

The ecosystem you live in

1 a Alexander von Humboldt, a German explorer, collected a great deal of data as he travelled the world. What did he find out about the number of species living in a place?

....

....

....

b Scientists have divided up the planet into biogeographical realms. What does the word 'realm' mean?

....

....

....

c What word did the German biologist Ernst Haeckel devise to describe the relationships between plants, animals and the environment?

....

A vocabulary of ecology

2 What is

a an ecosystem? ....

....

....

b a habitat? ....

....

....

Three examples of ecosystems

3

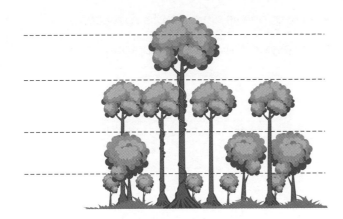

a Label the four layers in a rainforest on the diagram above.

b Name **three** organisms that you may find on the forest floor.

....

....

c Which layer of the rainforest is the habitat of monkeys?

....

d In which layer do bats roost?

....

4 Name **three** habitats found in the desert ecosystem. ....

....

5 The ocean ecosystem is divided into zones.

a Name **two** habitats in the intertidal zone. ....

....

b There are three zones in the open sea – the sunlit zone, the twilight zone and the dark zone. In which zone are the producers of food? Explain your answer.

....

....

Modelling an ecosystem

6 The diagram below shows some features and relationships in a simple ecosystem.

Write down the name of each feature of the ecosystem next to the appropriate letter.

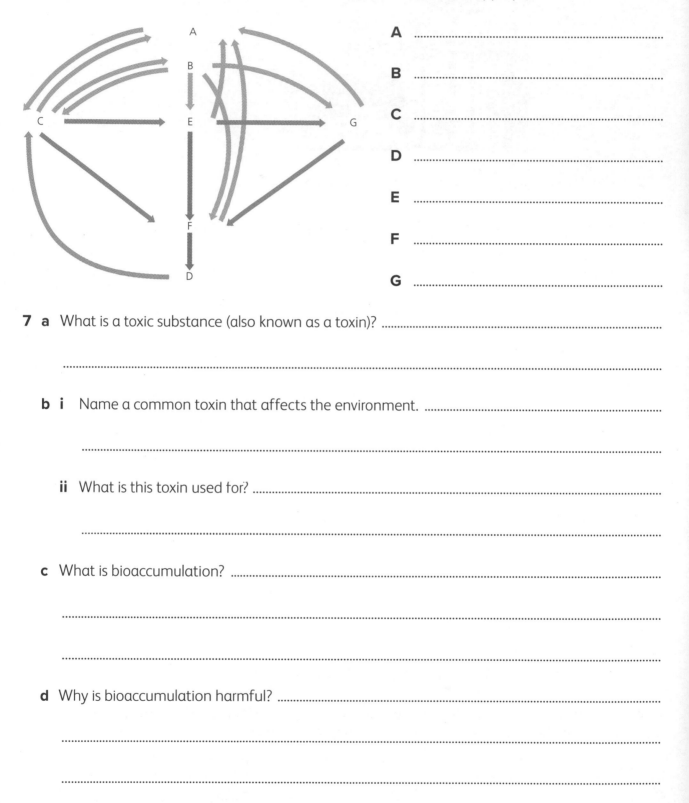

A ....

B ....

C ....

D ....

E ....

F ....

G ....

7 a What is a toxic substance (also known as a toxin)? ....

....

b i Name a common toxin that affects the environment. ....

....

ii What is this toxin used for? ....

....

c What is bioaccumulation? ....

....

....

d Why is bioaccumulation harmful? ....

....

....

8 a What substance did Paul Muller discover that could kill insects and was cheap to make? State the three letters it is known by.

....

b This substance was sprayed on Lake Clear in North America. What type of insect did it kill?

....

c This substance went into the plankton in the lake and then what happened to it?

....

....

....

9 a What is an invasive species? ....

....

....

b Name **two** invasive species in the ecosystems of Australia.

....

....

c How does an invasive species of plant affect the other plants that occur naturally in an ecosystem?

....

....

....

10 Crayfish are freshwater crustaceans that look like small lobsters. In the river and stream ecosystems of one country, the white-clawed crayfish lives naturally. It grows to about 12 cm in length and feeds on small insects and molluscs. Young crayfish shelter in the riverbanks under rocks and roots, while adults move out into the running water and shelter under rocks and logs.

The signal crayfish, which grows to a length of up to 18 cm, was brought into the country to be raised for food, but it escaped and spread into the rivers and streams. Signal crayfish eat the same food as the native white-clawed crayfish and hide in the same places, but also burrow into the riverbanks and break them down, releasing soil into the water which fills the spaces between rocks and logs. The signal crayfish also carries a disease to which it is immune, but which is deadly to the white-clawed crayfish.

As the numbers of signal crayfish increased, the numbers of white-clawed crayfish decreased. Give reasons why this change may have occurred.

....

....

....

....

....

....

....

11 An ecologist studying a woodland regularly sees insects of beetle species A running over the ground between the stones and plants. Occasionally she sees a few insects of beetle species B and wonders if they are an invasive species. What suggestions could you make to her for testing her idea?

7 Investigating an ecosystem

Investigating an ecosystem

1 State **two** reasons to set up habitat surveys.

1 ....

....

2 ....

....

2 When scientists make a survey, they may assess the numbers of a species in a habitat and use the DAFOR code. D stands for 'dominant' meaning that the species is found in large numbers in the habitat, but what do the other letters stand for?

A ....

F ....

O ....

R ....

Making a survey

3 Ben and Aya plan to make a survey of a wood called Greenwood, on the edge of a town. They are given some secondary sources to consider. Here are the titles of the sources.

A The plants and animals of Greenwood

B Greenwood would be great for new houses

C Seaweeds of the sea shore

D Greenwood is only a place for collecting wood for a fuel to burn

E Artists find much to draw and paint in Greenwood

F Plants that grow on mountain tops

a Which titles do you think may be biased in some way?

....

b Which titles are irrelevant?

....

c Which titles might provide useful information when making the survey?

....

4 Here is a list of equipment used in making habitat surveys.

A pitfall trap   **C** Tullgren funnel   **E** sweep net

B transect rope   **D** quadrat   **F** sheet and beater

a Which pieces of equipment are used to survey plants in a habitat?

....

b Which pieces of equipment are used to survey animals close to the ground?

....

c Which piece of equipment is used to survey animals in long grass?

....

5 A lawn looks to be formed entirely of grass, but when a small piece of it is examined closely, there are small yellow flowers growing in it. You are asked to make a survey of the lawn to find out how frequently the yellow flowers are found to be growing among the grass.

a What equipment will you use to make the survey?

....

b How will you use this equipment?

....

....

c How will you make your data reliable?

....

....

d How will your conclusion be limited? How could the survey be improved?

....

....

6 A beetle has been observed moving around on the ground of a woodland habitat. Some people say that it is rare in the habitat and others say that it is common. Make a plan to survey the ground by

a selecting a number of pieces of equipment to provide reliable data

....

b describing how the equipment will be set out to provide reliable data

....

....

c describing how you will use the equipment to provide reliable data

....

....

d describing how you will draw a conclusion

....

....

e explaining any limitations of the conclusion.

....

....

f What improvements could be made?

....

....

....

7 What should you consider when making a risk assessment about investigating a habitat, and how could the risks be controlled?

....

....

....

....

....

....

8 You have been asked to find out how fast a new lawn of grass is invaded by other species, by using a square frame called a quadrat.

a Write a plan for an enquiry which takes place over an extended time period to find out how fast the grass ecosystem is invaded.

....

....

....

....

....

....

....

b If there are several invasive species, how could you find out which one is the most invasive?

....

....

....

8 The structure of atoms

1 Democritus thought that if you cut something up into smaller and smaller pieces you would come to a particle which he called an atom. What is a property of this atom? Tick (✔) **one** box.

It is invisible. ☐

It is invincible. ☐

It is indivisible. ☐

It is indescribable. ☐

2 Aristotle thought that matter was made from four things – water and air were two of them. What were the other two?

....

3 Antoine Lavoisier studied chemical reactions and set out a law called the Law of conservation of mass. Here is a statement about it. Fill in the words that are missing.

Matter is neither .... nor .... in a chemical reaction.

4 Joseph Proust studied chemical reactions and devised the law of definite proportions. Here is a statement about it. Fill in the words that are missing.

.... in a .... are always present in a certain

definite proportion, no matter how the .... was made.

5 John Dalton set out his atomic theory. Here are some statements about atoms and elements. Which two are in his theory? Tick (✔) **two** boxes.

Atoms of an element have different masses and properties. ☐

Atoms of an element have the same mass and properties. ☐

Atoms of different elements have different masses and properties. ☐

Atoms of different elements have the same masses and properties. ☐

Ernest Rutherford and the atom

6 Ernest Rutherford worked with other scientists and discovered alpha particles, beta particles and gamma rays.

a Which of these three did he use to investigate the structure of the atom?

....

b Rutherford used a thin sheet of metal in his experiment. What was the metal?

....

c The first results from Rutherford's experiment suggested that the atom had a structure like which of the following? Tick (✔) **one** box.

A mango ☐   A plum pudding ☐

A chicken pie ☐   A bag of rice ☐

Electrons, protons and neutrons

7 The diagram shows the structure of the atom as we know it today. Label parts A, B and C on the diagram to the right.

8 Where in the atom are protons and neutrons found?

....

9 How are protons and neutrons

a similar? ....

b different? ....

The electrostatic force inside the atom

10 a What is the electrical charge on an electron? ....

b To which other subatomic particle is an electron attracted? ....

c What is the force that attracts the electron and this particle? ....

11 Fazil has a woollen jumper and holds up a balloon on a thread. He rubs the balloon on a woollen sleeve then holds it up by the thread again. He brings it close and the balloon moves.

Aya and Joel are watching and Fazil says the experiment shows that force has been generated.

Shazia believes that the experiment is incomplete. What should Fazil do to make his conclusion more reliable?

....

....

....

9 Mixtures and impurities

Pure and impure substances

1 What is

a a pure substance?

....

....

b an impure substance?

....

....

....

Detecting pure and impure substances

2 The table shows the temperatures at which four substances were seen to melt and boil. (These substances are not real, they have been created only for the purpose of this question.)

Substance	Temperature at which substance melted/°C	Temperature at which substance boiled/°C
A	10–15	40–50
B	65	180
C	129	204
D	90–97	310–314

Which of the substances are pure? Explain your answer.

....

....

Causes of impurity

3 Here is the general word equation for a chemical reaction:

Reactant A + Reactant B $\rightarrow$ Product C + Product D

   a In a chemical reaction the products C and D also have reactant A with them as an impurity.

      What has happened to produce this result?

      ....

      ....

   b In a chemical reaction, only small amounts of product C and D are produced, and both reactants A and B are present as impurities.

      What has happened to produce this result?

      ....

   c In a chemical reaction, the products also have substance X present as an impurity.

      Where can X have come from?

      ....

4 Sam puts a spatula full of copper carbonate in an empty beaker. He carefully adds some sulphuric acid and watches it fizz.

   When the fizzing stops, he says that he has got the maximum quantities of two pure products in the beaker.

   Aya says that he is wrong.

   Who is correct? Explain your answer.

   ....

   ....

   ....

   ....

5 a Name **two** dangerous properties of sulfuric acid to consider when making a risk assessment.

....

b What advice would you give in a risk assessment for a person using acids and pouring them from beakers onto reactants.

....

....

....

Dissolving and concentration

6 a When scientists study solubility they use the terms 'solvent', 'solute' and 'solution'. What do these terms mean?

i Solvent ....

....

....

ii Solute ....

....

....

iii Solution ....

....

....

b Scientists also use the word 'concentration' when studying solubility. What does this word mean when used scientifically?

....

c What is the difference between a solution with a high concentration of a solute and a solution with a low concentration of a solute?

....

....

7 a In simple terms, what does the solubility of a substance mean?

....

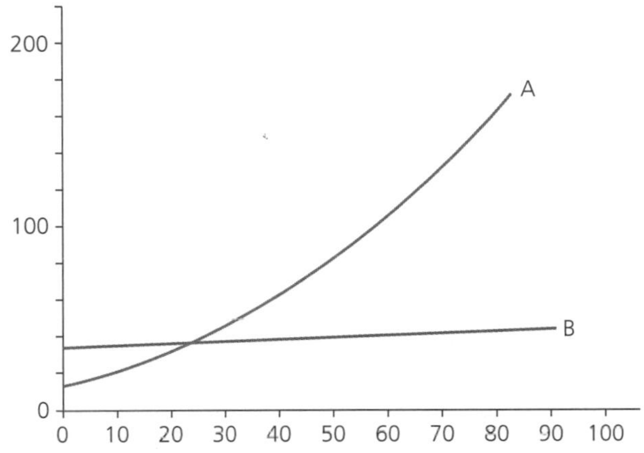

The graph above shows the solubility of two substances.

b Label the X axis and the Y axis.

c Does the graph show a trend in solubility? Explain your answer.

....

....

d How does the solubility of A compare with the solubility of B?

....

....

Chromatography

 8 Alex puts a drop of ink on a line on a filter paper strip and dips the strip in a solvent. After half an hour, his experiment looks like the diagram below.

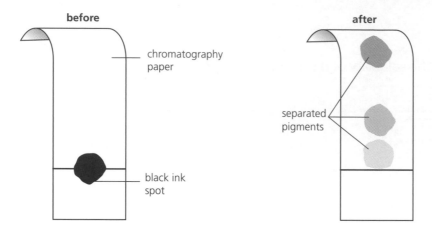

a What has happened to the solvent during the half-hour?

....

b Which pigment in the ink has moved fastest up the paper?

....

c State **two** reasons why one pigment moves faster than the other.

....

....

9 a What type of chromatography is used to investigate the contents of food?

....

b Chromatography is used to find acids in foods. What can these acids do to foods?

....

c Chromatography is used to test food for additives. What do the additives do to food?

....

d If additives are not added to food, what gives them their taste?

....

e i What type of chromatography is used in forensic science?

....

ii Name **two** types of samples from a crime scene that are tested by this type of chromatography.

....

f i How is chromatography used in studies of the environment?

....

ii State **two** uses of chromatography in medicine.

....

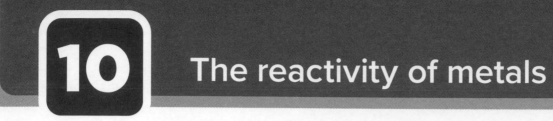

1 a How is sodium stored? ....

   b Explain the reason for this method of storage. ....

      ....

Reaction of metals with oxygen

2 Use the following words and phrases in your answers to the parts of this question.

   - black powder
   - glows
   - not changed
   - makes yellow sparks
   - forms black powder on the surface

   What happens when each of the following metals is heated?

   a Iron ....

   b Gold ....

   c Copper ....

Reaction of metals with acids

3 Jaafar is selecting apparatus to investigate the reaction of some metals with an acid.

   a Which of the following pieces of apparatus should he choose? **Circle** your answers.

   - trough
   - thermometer
   - Bunsen burner
   - delivery tube
   - thistle funnel
   - bung with two holes

   - bung with one hole
   - beaker
   - test tube
   - tripod
   - gauze
   - conical flask

   After he has assembled the apparatus, put in the metal and added the acid, there is a fizzing and a gas is produced.

   b i What is this gas? ....

      ii Where is it collected? ....

Jaafar has used hydrochloric acid in his investigation and has begun to write the word equation for the reaction he observed.

c Complete the following word equation:

metal + hydrochloric acid → ....

d Jaafar decides to set up the apparatus again and try a second metal. How can he find out whether this second metal was more or less reactive than the first metal he investigated?

....

....

....

e If he tested the following metals, which ones would he see react with the acid? **Circle** your answers.

– silver
– copper
– magnesium

– calcium
– gold
– zinc

The reactivity series

4 Here are some metals that are in the reactivity series. List the top five in the correct order.

– copper

potassium
– calcium

– zinc
– iron
– gold
– sodium

– magnesium

1 ....

2 ....

3 ....

4 ....

5 ....

5 a What does it mean when a substance is described as 'inert'?

....

b Name **two** metals in the reactivity series that are inert.

....

Exothermic and endothermic reactions

Exothermic reactions

1 What is an exothermic reaction?

....

....

2 Which of these words describes a chemical reaction in which a substance reacts with oxygen and heat is given out? Tick (✔) **one** box.

combination ☐

condensation ☐

combustion ☐

concentration ☐

3 a What is a fuel? ....

....

b Name **three** examples of fuels. ....

....

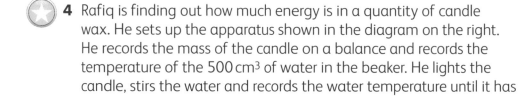

4 Rafiq is finding out how much energy is in a quantity of candle wax. He sets up the apparatus shown in the diagram on the right. He records the mass of the candle on a balance and records the temperature of the 500 cm³ of water in the beaker. He lights the candle, stirs the water and records the water temperature until it has risen by 10 °C. He then puts out the burning candle and records its mass again.

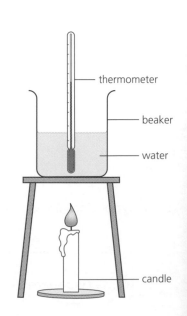

thermometer

beaker

water

candle

Here are Rafiq's results.

Mass of candle before burning	10 g
Mass of candle after burning	5 g
Loss of mass	

Temperature before heating	22 °C
Temperature after heating	32 °C
Rise in temperature	

a Complete the two tables.

The energy in 1 g of candle is found by using the formula below:

$$\frac{2.1 \times \text{rise in temperature}}{\text{loss of mass}} = \text{kJ/g}$$

b Use the formula to find the energy in the candle wax that was burnt away by Rafiq.

5 Jaya is comparing the heat produced by two fuels. She uses each one in turn to heat an equal volume of water for 10 minutes, and records the temperature every 2 minutes. The table below shows her data.

Time/minutes	Fuel A temperature/°C	Fuel B temperature/°C
0	20	20
2	22	24
4	24	28
6	26	32
8	28	36
10	30	40

a Draw and label a line graph for each fuel.

b Use the graph to predict what the temperature of each beaker of water would have been after 12 minutes. Mark your predictions with a dot.

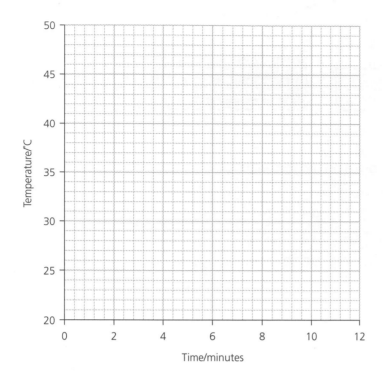

c Using this predicted data, calculate the temperature difference between the two beakers of water after 12 minutes.

....

....

6 a What kind of calorimeter is used to measure the energy in food?

....

b What gas is used in the calorimeter to burn the food?

....

7 The diagram below shows a two pot cooking stove from Sri Lanka.

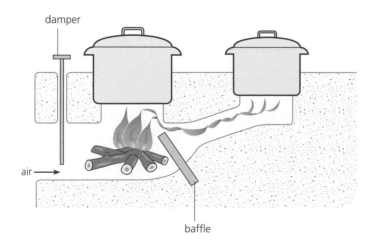

a Which of these statements describes the function of the stove? Tick (✔) **two** boxes.

It uses two lots of fuel to heat two pots. ☐

It uses one lot of fuel to heat one pot. ☐

It uses one lot of fuel to heat two pots. ☐

It helps to conserve fuel in the region. ☐

b When the damper is raised what happens to

i the amount of air reaching the fuel? ....

....

ii the way the fuel burns? Explain your answer. ....

....

....

....

c How is the heat that is cooking the large pot reduced? Explain your answer.

....

....

d What happens when the baffle is removed to open the pipe to the second pot?

....

8 Draw lines to match each part of a hand warmer to its function.

Part	Function
iron powder	catalyst (speeds up reaction)
charcoal	insulation (slows down the release of heat)
salt water	spreads out heat
vermiculite	releases heat in an exothermic reaction

49

Endothermic reactions

9 What is an endothermic reaction?

....

10 a Heat is taken in when ice melts. Is melting an endothermic chemical reaction?

....

b Explain your answer. ....

....

....

11 a Rearrange these substances to make a word equation for the reaction that occurs when you put a sherbet sweet in your mouth.

- water
- citric acid
- carbon dioxide
- sodium citrate
- sodium hydrogen carbonate

....

....

b What substance makes the sherbet fizz?

....

c How does it feel when you put sherbet in your mouth?

....

d What does this sensation tell you about the chemical reaction taking place?

....

12 What kind of chemical reaction takes place when food is cooking in a hot pan?

....

13 a What type of injuries are ice packs used to treat?

....

b An ice pack contains a bag of a nitrogen compound called ammonium nitrate. It also contains another bag. What is in this second bag?

....

c What is done to the ice pack to make it useful to treat an injury?

....

....

d What kind of chemical reaction takes place inside the ice pack when it is brought into use?

....

e Name **two** ways in which using an ice pack helps treat the injury.

....

....

The speed equation

1 What is speed? Rearrange these phrases to make a definition.

- – in a certain time
- – a moving object
- – the distance covered by
- – a measure of

Speed is ....

....

2 Write down the speed equation.

....

3 a A friend wants to know her speed as she travels along pedalling steadily on her bicycle. You have two poles that you can stick in the ground, a sports tape measure that is 50 metres long and a stop watch. How could you record her speed?

....

....

....

....

....

....

b What useful risk assessment would you make before making this investigation?

....

....

....

....

Investigating speed

4 Peter challenges three other members of his class to a one-lap race. Usen, Jamil, Sari and Eleanor stand by the start/finish line and use stop-watches to record when the runners set off and when they complete their lap.

a State **two** ways in which errors may occur in recording the runners' speeds.

....

....

b How could the recording of the runners' speeds be made much more accurate?

....

....

5 a What is a light gate? Explain using the terms 'light beam' and 'light sensitive switch'.

....

....

....

b Describe how you would use two light gates to measure the speed of a toy car moving down a ramp.

....

....

....

....

c How would you collect reliable data about the speed of the car down the ramp?

....

....

d If you repeated the experiment by making the ramp steeper and then even steeper again, would you expect the three sets of data to show a trend? Explain your answer.

....

....

6 a What does a radar gun fire? ....

b How is reflection used in the process of finding the speed of a vehicle?

....

....

c What does the computer in the radar gun do to calculate the speed of a vehicle?

....

....

Distance/time graphs

7 Nadia made the following observations on an insect. It moved 1 cm in 10 seconds, then 2 cm in the next 10 seconds; it stayed still for 10 seconds, then moved 5 cm in the next 10 seconds; it stayed still for 20 seconds, then moved 2 cm in the next 30 seconds.

a Draw a distance/time graph of the insect's movement.

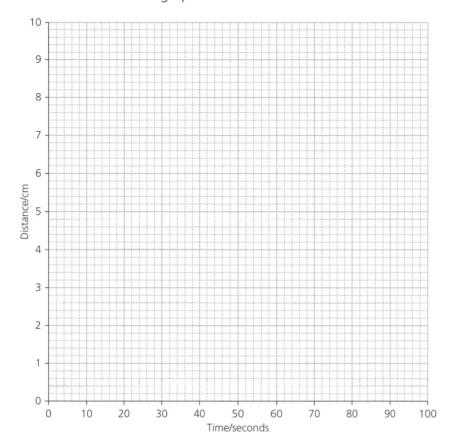

b When did the insect travel fastest? ....

c When did the insect travel slowest? ....

d How far would the insect have got in 1 minute if it had kept moving at its original speed?

....

e Nadia then saw the insect move 20 cm in 2 seconds. What do you think it did?

....

....

....

13   Forces

1 The following events happened in a football game.

Abu kicks a ball (A). He runs after it and kicks it again before it stops (B). Carla kicks the ball back (C) to Abu, who traps the ball with his foot (D) and squashes it (E). Fill in the table by matching each action letter in the game with the action of a force.

Action of a force	Action in the game
changing the direction of a moving object	
stopping a moving object	
changing an object's speed	
changing an object's shape	
starting an object moving	

2 The force which supports you on the ground by pushing up on you from the ground is called the normal force.

   a What is the force that pushes your body down against the ground?

   ....

   b Why do you not sink into the ground?

   ....

   ....

3 a In the space on the next page, draw a car that is parked on a road. Draw two force arrows to show the balanced forces acting on the car.

b When the car is moving at constant speed there is another force which is being produced by the engine. What is it?

....

....

c When the car moves forwards, another force acts upon it which is produced by the air. What is it?

....

d In the space provided, draw the four forces acting on a car travelling at constant speed.

4 a What is the force pushing down on a boat floating on a lake?

....

b What is the force pushing up on the boat into the water?

....

c Why does the boat not sink? ....

....

....

5 Which two statements about plimsoll lines are true? Tick (✔) **two** boxes.

Plimsoll lines show the maximum level to which the water should rise on the side of a ship when it is being loaded in a dock. ☐

Plimsoll lines show the minimum level to which the water should rise on the side of a ship when it is being loaded in a dock. ☐

An over-loaded ship will never sink in a storm. ☐

Warm and cold waters in the world's oceans push up on ships with different forces, which can affect the ships' ability to stay afloat. ☐

Balanced forces

6 a When you start moving a go-kart forward you press on a pedal. What is the pedal called?

....

b As you move forwards what is

i the force that moves the go-kart?

....

ii the force that pushes on the kart against this force?

....

c If the go-kart goes faster and faster, are these two forces balanced or unbalanced? Explain your answer.

....

....

....

d What happens to the force produced by the engine when you take your foot off the pedal you have been pushing?

....

e There is a second pedal that you push that makes the kart slow down quickly and stop. What is the name of this pedal?

....

f When you press this pedal, are the forces on the go-kart balanced or unbalanced? Explain your answer.

....

....

....

Unbalanced forces

7 a When a boat is sinking, are balanced or unbalanced forces acting on it? Explain your answer.

....

....

b Why does a boat sink when it has a hole in it?

....

....

8 a What is a bathyscaphe used for?

....

b If you were in a bathyscaphe and wanted it to sink, what would you do and why?

....

....

c If you were in a bathyscaphe and wanted to rise in the water, what would you do and why?

....

....

Turning forces

9 a What is the moment of a force?

....

....

b Write down the equation for the moment of a force.

....

10 Shen found a long plank and set it up as a seesaw. His weight is 450 N and he sits 2 metres from the point on which it turns. How far will his younger sister, Bo, who weighs 300 N, have to sit from the point in order to balance her brother? Show your working.

....

....

11

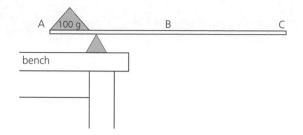

a Which of these predictions would you select to make the beam shown above horizontal?

   The beam can be made horizontal by

   i   using the same mass at both A and B.

   ii keeping the same mass A but placing a smaller mass at B.

   iii keeping same mass at A and a mass at C which is the same mass as that at A.

   ....

b How could you collect reliable data to test your prediction?

   ....

c How could you develop your experiment in order to test a trend in the masses you use on the long arm of the beam?

   ....

   ....

   ....

   ....

Pressure on a surface

1 What is the equation used to define pressure?

....

2 A block has three surfaces labelled A, B and C.

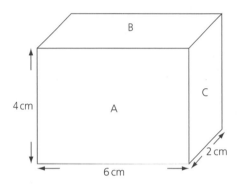

a What is the area of each of the three surfaces? Show your working.

A ....

....

B ....

....

C ....

....

b The weight of the block is 48 N. What is the pressure on the ground when the block is stood on each surface? Show your working.

A ....

....

B ....

....

c ....

....

3 Two identical cows fell into a muddy hole. One fell on its side and the other fell on its feet.

a Which one sank furthest into the mud? ....

b Explain your answer. ....

....

Reducing pressure

4 Ingrid lives in a cold country that sometimes gets lots of snow. When that happens, people can enjoy the sport of skiing. Ingrid goes outside to see if the snow is fit for skiing. She sinks into the snow, but when she puts her skis on, she can move over it without sinking. Why?

....

....

Increasing pressure

5 Paulo is playing football in his trainers but he keeps slipping. He changes into boots with studs and stops slipping.

Why? ....

....

6 A drawing pin (thumb tack) has a head and a point. When you push it into a board, which part is under

a high pressure? ....

b low pressure? ....

7 A chef is having difficulty cutting up onions because his knife is blunt. After he sharpens it, the knife cuts more easily.

Why is this? ....

....

....

Particles and pressure

8 A beaker contains a liquid. State **two** places where the liquid exerts pressure.

....

9 Use the particle theory to explain how the pressure of water on the bottom of a beaker changes when you pour more water into it.

....

....

....

....

10 The picture below shows three jets of water flowing from a can.

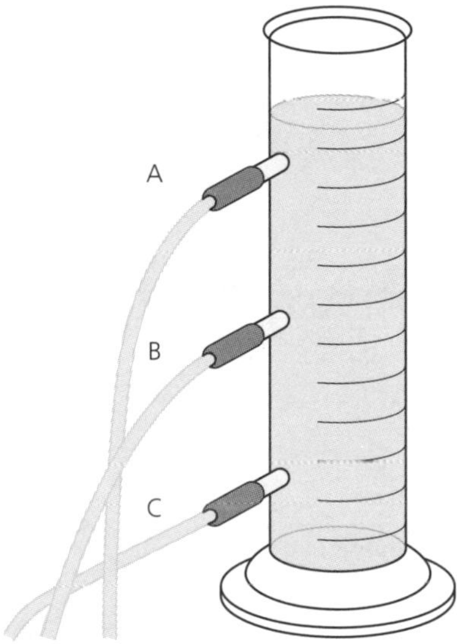

a Which jet of water is under the greatest pressure and which is under the least pressure?

....

b How can you tell?

....

c Use the particle theory to explain the difference in pressure.

....

....

11 a Imagine you are an engineer and your project is to build a dam to create large reservoir in a valley. Which of these plans would you choose? Tick (✔) **one** box.

Make the dam wall thicker at the top and thinner at the bottom. ☐

Make the dam wall thinner at the top and thicker at the bottom. ☐

Make the dam wall thicker in the middle and thinner at the top and bottom. ☐

Make the dam wall the same thickness from the top to the bottom. ☐

b Explain your choice from part a.

....

....

....

12 a What is hydraulic equipment used for? Tick (✔) **one** box.

Transmitting liquids from one place to another. ☐

Transmitting pressure from one place to another. ☐

Transmitting pressure in all directions. ☐

Transmitting pressure from one liquid to another. ☐

b Name **two** uses of hydraulic systems. ....

....

13 a What is a hovercraft used for?

....

b What advantage does it have over a boat or a road vehicle?

....

....

c The hovercraft has a cushion of air beneath it. What does this cushion of air do to the hovercraft?

....

....

Diffusion

14 Draw three diagrams to show how particles of a scent in just one place diffuse into the air around them.

15 Light

Reflection of light

1

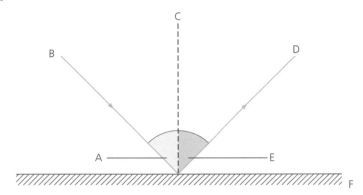

a What are the labels for the following letters?

A ....

B ....

C ....

D ....

E ....

F ....

b How would you describe the item labelled C? ....

2 Which of these statements is the law of reflection? Tick (✔) **one** box.

When a ray of light strikes a mirror, its angle of reflection is the same as the angle
of incidence. ☐

When a ray of light strikes a mirror, its angle of reflection is less than the angle of incidence. ☐

When a ray of light strikes a mirror, its angle of reflection is more than the angle
of incidence. ☐

When a ray of light strikes a mirror, its angle of reflection is never the same as the angle
of incidence. ☐

3 a On what kind of surface might you see an image?

....

b What produces the image you see?

....

4 a What did the Ancient Greeks believe to be the speed of light?

....

b Ole Romer, a Danish astronomer, studied the moons of a planet to find the speed of light. Which planet?

....

c Which of the following best describes the speed of light? Tick (✔) **one** box.

30 m/s ☐

300 000 000 000 m/s ☐

30 000 ms ☐

300 000 000 ms ☐

d If the Sun suddenly stopped shining, roughly how long would it take for the Earth to go dark?

....

Refraction of light

5

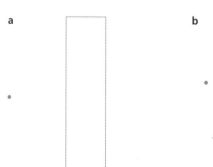

This diagram shows two transparent glass blocks. The dot to the left of each one represents a light source.

a On the left hand block, draw a line to show the path of a ray of light from the source to the other side of the block.

b On the right hand block, draw a line to show the path of a ray of light from the source to the other side of the block.

6

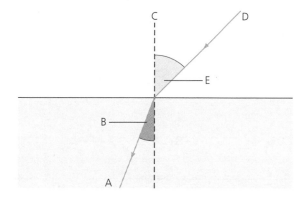

What are the labels for the following letters?

A ....

B ....

C ....

D ....

E ....

7 a If light slows down as it passes from one transparent material to another, how does the path of the light-ray change compared to the normal?

....

b If light speeds up as it passes from one transparent material to another, how does the path of the light-ray change compared to the normal?

....

The prism

8 a What is a prism?

....

b What is dispersion?

....

c What happens when white light strikes the side of a prism, travels through the prism and shines onto a screen on the other side?

....

....

....

....

9 a Who experimented with white light and two prisms?

....

b Why did he do the experiment?

....

....

c What did he do in his experiment?

....

....

....

d What observation did he make?

....

....

e What conclusion did he draw?

....

10 When the Sun shines on a blue flower, which colours are absorbed by its petals? **Circle** your answers.

- red
- orange
- yellow
- green

- blue
- indigo
- violet

11 What colours are reflected by

a the white paint on a car?

....

b the black tyres?

....

12 When white light shines on a red filter, some colours are subtracted from the light that leaves the filter.

a What does 'subtracted' mean?

....

b What happens in the subtraction process?

....

c Which colour is not subtracted from the white light?

....

13 a When two or three primary colours of light are combined together, what is the process called?

....

b What colour is produced when

i blue and green combine?

....

ii blue and red combine?

....

iii red and green combine?

....

14 a What are the tiny particles in paint that give the paint its colour?

....

b What colours of paint do you mix to get green?

....

c What happens if you mix the three primary colours of paint? Explain your answer.

....

....

....

....

The magnetic field

1 What is a magnetic field? Tick (✔) **one** box.

a A region around a magnet in which the push of the magnetic force acts on magnetic materials. ☐

b A field in the countryside where a magnet points north and south. ☐

c A region around a magnet in which the pull of the magnetic force acts on magnetic materials. ☐

d A region around a magnet where only other magnets can exert a force. ☐

2 a The diagram below shows the outline of a bar magnet that is under a card. Draw what would happen if iron filings were sprinkled on the card above the magnet.

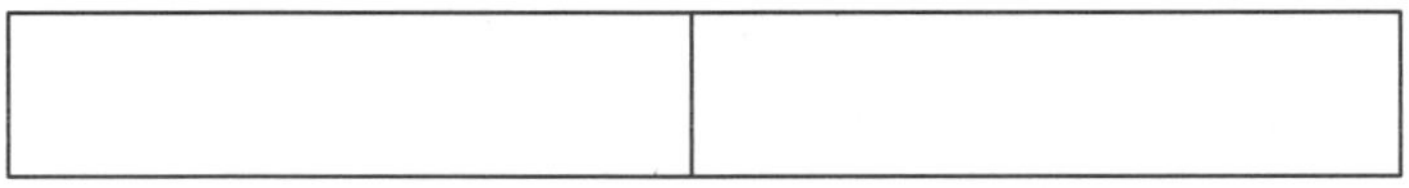

b Why do iron filings behave in this way?

....

....

c If one end of a bar magnet is dipped into a pot of iron filings and raised up, the iron filings form spiky columns in all directions. Why do they do this?

....

....

d i Iron filings are small metal particles that could get in the eyes if not handled safely. How could someone use a sealable, clear plastic bag to protect their eyes when looking at how iron filings behave when brought near a magnet? Explain your answer.

....

....

....

ii Suggest another precaution to take to prevent iron filings from getting into the eyes.

....

....

3 a What sort of a compass is used to investigate magnetic fields?

....

b Here are some steps in using the compass to investigate a magnetic field, but they are in the wrong order.

Write down the letters of the steps in the correct order.

A Take a pencil and mark a point on the paper where the needle is pointing.

B Place the compass close to one pole of the magnet and note the direction in which the needle points.

C Take the pencil again and mark the point on the paper where the needle is pointing.

D Place a bar magnet in the centre of a card.

E Move the compass just ahead of the point you have just made and note the direction the needle is pointing now.

....

Magnetic field patterns

4 a In the space on the next page

i draw two bar magnets with the south pole of one pointing towards the north pole of the other.

ii draw in the magnetic field pattern of the two magnets.

b In the space below

   i draw two bar magnets with the north pole of one pointing towards the north pole of the other.

   ii draw in the magnetic field pattern of the two magnets.

5 The diagram on the next page shows the Earth and its geographic north and south poles.

   a On the diagram, mark the position of the magnetic north pole and draw in the magnetic field pattern around the planet.

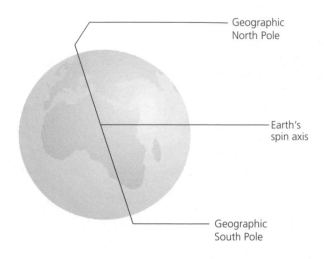

Geographic
North Pole

Earth's
spin axis

Geographic
South Pole

b Imagine that the mark you have drawn represents the position of the magnetic north pole today. If scientists checked its position in a few years' time, would it still be in the same place? Explain your answer.

....

....

c Which part inside the Earth acts like a magnet?

....

6 a Which civilisation was probably the first to use a magnet?

....

b What did they use it for?

....

....

c Who was the scientist in this civilisation that first used the magnet scientifically?

....

d How did his studies help people separated by seas to trade with each other?

....

....

e How did Peregrinus change the design of the compass?

....

....

f Name **two** beliefs about magnets that William Gilbert disproved in his experiments.

....

....

g Gilbert made a model to help him understand how compasses behaved on the surface of the Earth.

   i What shape was the model?

....

   ii What material was the model made from?

....

   iii What was the model an analogy of?

....

   iv What did Gilbert discover by experimenting with the model?

....

....

The link between magnetism and electricity

7 What happens when a compass is placed near a wire and a current of electricity is passed through the wire?

....

....

....

8 What happens when magnetic lines of force come closer together? Tick (✔) **one** box.

The magnet reverses its poles. ☐

The magnetic force decreases. ☐

The magnetic force does not change. ☐

The magnetic force increases. ☐

The electromagnet

9 How would you make an electromagnet from a nail and a length of copper wire?

....

....

....

....

10 Can you test an electromagnet using a clamp and stand, steel paper clips, a cell, three wires and a switch?

   a Make a drawing of the experimental set up or describe it.

   b What would you do to see if you had made an electromagnet?

....

....

....

11 The strength of an electromagnet was tested by finding out how many paperclips it could pick up when currents of different sizes were passed through it.

Here is the data that was collected.

Current (A)	Number of paper clips
0	0
1	10
2	25
3	30
4	40

a On the grid below, plot a graph using the data in the table.

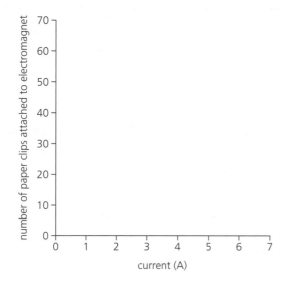

b Is there an anomalous result? Explain your answer.

....

....

c Does the data show a trend? Explain your answer.

....

d How many paper clips do you predict will stick to the electromagnet when a current of 5 amps passes through it?

....

e How would you make the data more reliable?

....

....

....

17 Using the Earth's resources

Renewable and Non-renewable material resources

1 Name **two** resources that humans have needed to use since the earliest times.

....

....

2 Why is wood a renewable material?

....

....

....

3 Name **four** renewable materials used to make bioplastics.

....

....

4 Why is it claimed that bioplastics are better for the environment than normal plastics made from fossil fuels?

....

....

....

5 Are metal ores and minerals renewable materials? Explain your answer.

....

....

....

Renewable energy resources

6 Is tidal power a renewable or a non-renewable form of energy? Explain your answer.

....

....

....

....

7 You are to make a windmill (turbine) with paper blades for use in an experiment. When you have made your windmill, you should test it to see if the blades turn smoothly.

   a How will you test your windmill?

....

....

   b How will you set up an experiment to see if the length of the blades affects the way they turn.

....

....

....

....

   c From your knowledge of materials such as paper, make a prediction of what you may expect to find as the blades become longer.

....

....

....

   d Is a risk assessment needed when using turbine blades? Explain your answer.

....

....

8 a The wind is made from moving particles of air. What form of energy do they have?

....

b In a wind turbine, this energy is transferred into another form that we use in a variety of ways. What is this form of energy?

....

c State two places where turbines should not be sited because of air turbulence.

....

d State one place that turbines could be sited as the wind flows more smoothly.

....

e State one benefit to the environment of using wind turbines.

....

....

f State two ways in which some people think turbines harm their environment.

....

....

9 a Name **two** objects in space whose gravitational pull makes the tides.

....

b Water at high tide and low tide has energy stored in it. Which energy resource is used for generating electricity?

....

10 a Where does solar power come from?

....

b What form of energy do solar panels trap?

....

c What form of energy is converted to electrical energy in a solar cell?

....

Non-renewable energy resources

11 a Give **two** examples of fossil fuels.

....

b Are fossil fuels a renewable or a non-renewable resource? Explain your answer.

....

....

....

....

....

....

....

12 The first people lived as hunter-gatherers.

a Name **three** material resources they would have used.

....

b Name **two** things they would have used these materials for.

....

c Name an energy source and what they used it for.

....

d When people began farming, it allowed other crafts to develop that needed material resources. Name **two** of these crafts.

....

e When steam engines were developed and used in the Industrial Revolution, what energy resource did they use?

....

18 The Earth's climate

Weather and climate

1 What is the difference between weather and climate?

....

....

....

....

2 a How could you use an orange and a sheet of cling film to model the Earth and its atmosphere?

....

b What does this tell you about the atmosphere?

....

....

The Earth's climate

3 a Where does the atmosphere get its heat energy from?

....

b Name **three** things that heat energy does.

....

....

4 Name **four** measurements that scientists make when they collect data in order to monitor the Earth's climate.

....

....

5 The climate of the Earth has had cold periods and warm periods.

   a What are the two names that are used for the cold periods?

   ....

   b What is the name given to a warm period?

   ....

6 Fossil coral is found in various levels of rocks, but after a certain level they are no longer discovered in rocks higher up.

   How can this information be used to investigate climate change?

   ....

   ....

   ....

   ....

7 Scientists use evidence from sedimentary rocks to help them find out about climate change in the past.

   Sedimentary rocks are produced when other rocks are broken down by the atmosphere, in a process called erosion.

   a What does the word 'erosion' mean?

   ....

   ....

   b How do the products of erosion settle down to make sedimentary rocks?

   ....

   c If a sedimentary rock has a large amount of iron oxide in it, what does this tell you about the climate when the rock formed?

   ....

8 a Name **two** places where pollen grains from the past can be found.

   ....

b How are the pollen grains found at a particular time useful in investigating climate change?

....

....

....

9 Why can the following be useful in investigating climate change when found in ice cores?

a Dust: ....

....

....

b Volcanic ash: ....

....

....

c Bubbles of atmospheric gas: ....

....

....

....

10 The diagram below shows a very simple model of how the temperature of the atmosphere can change.

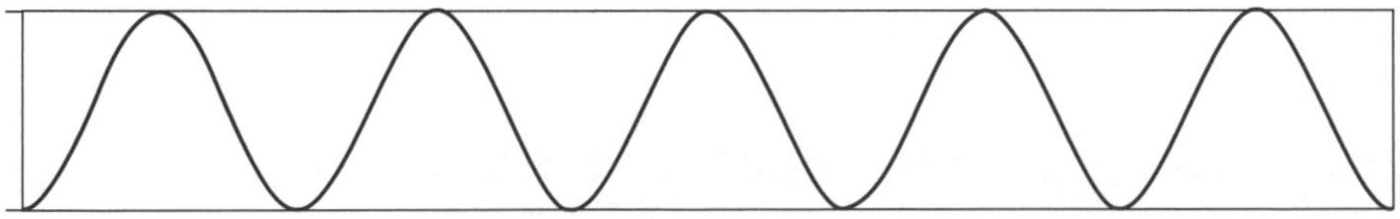

Does this model show a pattern or a trend? Explain your answer.

....

....

11 The graph below shows how the climate has changed over a hundred and sixty years. The global temperature anomaly is used to show how the temperature of the globe has changed compared to the zero (0) reading in the middle of the axis.

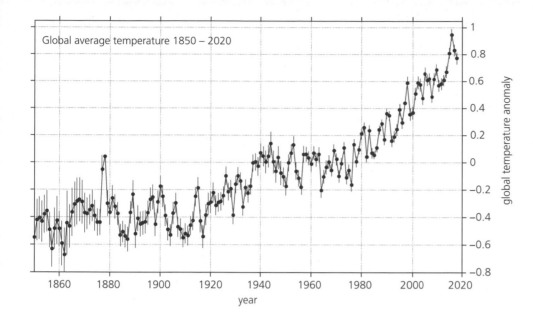

Does this graph show a pattern or a trend? Explain your answer.

....

....

19 Inside a galaxy

Galaxies

1 a What is the galaxy we live in?

....

b How many light years is it from one side to the other?

....

c How long does it take to go round once?

....

Intergalactic space

2 What is another name for stellar dust?

....

Planetary systems

3 a What is in a planetary system?

....

....

....

b What is our nearest planetary system?

....

Asteroids

4 a How does the size of an asteroid compare with the size of a planet?

....

b From what did asteroids form?

....

....

c Between which planets in the solar system would you find most asteroids?

....

d What is the region where most asteroids are found?

....

e Do the asteroids move in this region? Explain your answer.

....

The **Cambridge Checkpoint Lower Secondary Science** series consists of a Student's Book, Boost eBook, Workbook and Teacher's Guide with Boost Subscription for each stage.

Student's Book	Boost eBook	Workbook	Teacher's Guide with Boost subscription
Student's Book 7 9781398300187	eBook 7 9781398302136	Workbook 7 9781398301399	Teacher's Guide 7 9781398300750
Student's Book 8 9781398302099	eBook 8 9781398302174	Workbook 8 9781398301412	Teacher's Guide 8 9781398300767
Student's Book 9 9781398302181	eBook 9 9781398302228	Workbook 9 9781398301436	Teacher's Guide 9 9781398300774

The answers are FREE to download from:
www.hoddereducation.com/cambridgeextras
To explore the entire series,
visit **www.hoddereducation.com/cambridge-checkpoint-science**

Cambridge Checkpoint Lower Secondary Science Teacher's Guide with Boost subscription

Created with teachers and students in schools across the globe, Boost is the next generation in digital learning for schools, bringing quality content and new technology together in one interactive website.

The **Cambridge Checkpoint Lower Secondary Science Teacher's Guides** include a print handbook and a subscription to Boost, where you will find a range of online resources to support your teaching.

- **Confidently deliver the new curriculum framework:** Expert author guidance on the different approaches to learning, including developing scientific language and the skills required to think and work scientifically.

- **Develop key concepts and skills:** Suggested activities, knowledge tests and guidance on assessment, as well as ideas for supporting and extending students working at different levels.

- **Support the use of ESL:** Introductions and activities included that have been developed by an ESL specialist to help facilitate the most effective teaching in classrooms with mixed English abilities.

- **Enrich learning:** Audio versions of the glossary to help aid understanding, pronunciation and critical appreciation.

To purchase Cambridge Checkpoint Lower Secondary Science Teacher's Guide with Boost subscription, visit www.hoddereducation.com/cambridge-checkpoint-science

Cambridge checkpoint

Lower Secondary Science WORKBOOK

8

Practise and consolidate knowledge gained from the Student's Book with this write-in workbook full of corresponding learning activities.

- Save time when planning with ready-made homework or extension exercises.
- Reinforce students' understanding of key scientific concepts with varied question types and the use of ICT.
- Challenge students with extra practice activities to encourage regular self-assessment.

For over 30 years we have been trusted by Cambridge schools around the world to provide quality support for teaching and learning. For this reason we have been selected by Cambridge Assessment International Education as an official publisher of endorsed material for their syllabuses.

Working for over **30 YEARS** *WITH Cambridge Assessment International Education*

For more information on how to use this workbook, please visit: **www.hoddereducation.com/ workbook-info**

This resource is endorsed by Cambridge Assessment International Education

✓ Provides learner support as part of a set of resources for the Cambridge Lower Secondary Science curriculum framework (0893) from 2020

✓ Has passed Cambridge International's rigorous quality-assurance process

✓ Developed by subject experts

✓ For Cambridge schools worldwide

Boost
This series includes eBooks and teacher support.
Visit www.hoddereducation.com/boost for more information.

Registered Cambridge International Schools benefit from high-quality programmes, assessments and a wide range of support so that teachers can effectively deliver Cambridge Lower Secondary.

Visit **www.cambridgeinternational.org/ lowersecondary** to find out more.

HODDER EDUCATION
e: education@hachette.co.uk
w: hoddereducation.com

ISBN 978-1-398-30141-2

9 781398 301412

FSC
www.fsc.org

MIX
Paper | Supporting responsible forestry
FSC™ C104740